IEE 16th Edition Wiring Regulations

Inspection, Testing and Certification

By the same author

Electrical Installation Work

IEE 16th Edition Wiring Regulations: Design and Verification

Wiring Systems and Fault Finding for Installation Electricians

IEE 16th Edition Wiring Regulations
Inspection, Testing and Certification

Third Edition

Brian Scaddan IEng, MIIE (elec)

Newnes

OXFORD   AUCKLAND   BOSTON   JOHANNESBURG   MELBOURNE   NEW DELHI

Newnes
An imprint of Butterworth-Heinemann
Linacre House, Jordan Hill, Oxford OX2 8DP
225 Wildwood Avenue, Woburn, MA 01801-2041
A division of Reed Educational and Professional Publishing Ltd

 A member of the Reed Elsevier plc group

First published 1996
Second edition 1998
Third edition 2001

British Library Cataloguing in Publication Data
A catalogue record for this book is available from the British Library

ISBN 0 7506 5216 0

TK
3271
.S2692
2001

Printed on acid free paper
Composition by Genesis Typesetting, Laser Quay, Rochester, Kent
Printed and bound in Great Britain by Biddles Ltd, *www.biddles.co.uk*

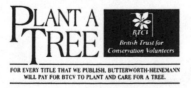

CONTENTS

PREFACE

As a bridge between the 16th Edition course (C&G 2380) and the Design, Erection and Verification course (C&G 2400), the author, in association with the City and Guilds and the NICEIC, developed the Inspection, Testing and Certification course (C&G 2391).

This book was written as an accompaniment to this new 2391 scheme. It is, however, a useful addition to the reference library of contracting electricians and candidates studying for the C&G 2380 and 2400 qualifications.

Brian Scaddan

INTRODUCTION

The IEE Wiring Regulations

Before we embark on the subject of inspection and testing, it is, perhaps, wise to refresh our memories with regards to one or two important topics from the 16th edition.

Clearly the protection of persons and livestock from shock and burns etc. and the prevention of damage to property, are priorities. In consequence, therefore, thorough inspection and testing of an installation and subsequent remedial work where necessary, will significantly reduce the risks.

So, let us start with electric shock, i.e. the passage of current through the body of such magnitude as to have significant harmful effects. Figure 1 illustrates the generally accepted effects of current passing through the human body.

How then are we at risk of electric shock, and how do we protect against it?

There are two ways in which we can be at risk:

1 Touching live parts of equipment or systems that are intended to be live, this is called *direct contact*.
2 Touching conductive parts which are not meant to be live, but have become live due to a fault, this is called *indirect contact*.

The conductive parts associated with indirect contact can either be metalwork of electrical equipment and accessories (Class 1) and that of electrical wiring systems such as conduit and trunking etc. called *exposed conductive parts*, or other metalwork such as pipes, radiators, girders etc. called *extraneous conductive parts*.

Introduction

1 mA – 2 mA	Barely perceptible, no harmful effects
5 mA – 10 mA	Throw off, painful sensation
10 mA – 15 mA	Muscular contraction, can't let go!
20 mA – 30 mA	Impaired breathing
50 mA and above	Ventricular fibrillation and death

Figure 1   Electric shock

Let us now consider how we may protect against electric shock from whatever source.

Protection against both direct and indirect contact

One method of protecting against shock from both types of contact relies on the fact that the system voltage must not exceed extra low (50 V AC, 120 V ripple free DC), and that all associated wiring etc. is separated from all other circuits of a higher voltage. Such a system is known as a *separated extra low voltage (SELV)*. If a SELV system exceeds 25 V AC, 60 V ripple free DC, then extra protection from *direct contact* must be provided by barriers, enclosures and insulation.

Another method of protection is by the *limitation of discharge of energy* whereby any equipment is so arranged that the current that can flow through the human body or livestock is limited to a safe level. Typical of this method is electric fences.

Protection against direct contact

Apart from SELV, how can we prevent danger to persons and livestock from contact with intentionally live parts? Clearly we must minimize the risk of such contact, and this may be achieved in one or more of the following ways:

1 Insulate any live parts.
2 Ensure that that any uninsulated live parts are housed in suitable enclosures and/or are behind barriers.
3 Place obstacles in the way. (This method would only be used in areas where skilled and/or authorized persons were involved.)
4 Placing live parts out of reach. (Once again, only used in special circumstances, e.g. live rails of overhead travelling cranes.)

The use of a residual current device (RCD) cannot prevent direct contact, but it may be used to supplement any of the other measures taken, provided that it is rated at 30 mA or less and has a tripping time of not more than 40 ms at an operating current of 150 mA.

It should be noted that RCDs are not the panacea for all electrical ills, they can malfunction, but they are a valid and effective back-up to the other methods. They must not be used as the sole means of protection.

Protection against indirect contact

How can we protect against shock from contact with live, exposed or extraneous conductive parts whilst touching earth, or from contact between live exposed and/or extraneous conductive parts? The most common method is by *earthed equipotential bonding and automatic disconnection of supply (EEBADS)*.

All extraneous conductive parts are joined together with a main equipotential bonding conductor and connected to the main earthing terminal, and all exposed conductive parts are connected to the main earthing terminal by the circuit protective conductors. Add to this overcurrent protection that will operate fast enough when a fault occurs and the risk of severe electric shock is significantly reduced.

Other means of protecting against indirect contact may be used, but are less common and some require very strict supervision.

Use of class 2 equipment

Often referred to as double-insulated equipment, this is typical of modern appliances where there is no provision for the connection of a CPC. This does not mean that there should be no exposed conductive parts and that the casing of equipment should be of an insulating material; it simply indicates that live parts are so well insulated that faults from live to conductive parts cannot occur.

Non-conducting location

This is basically an area in which the floor, walls and ceiling are all insulated. Within such an area there must be no protective conductors, and socket outlets will have no earthing connections.

It must not be possible simultaneously to touch two exposed conductive parts, or an exposed conductive part and an extraneous conductive part. This requirement clearly prevents shock current passing through a person in the event of an earth fault, and the insulated construction prevents shock current passing to earth.

Earth-free local equipotential bonding

This is in essence a Faraday cage, where all metal is bonded together but *not* to earth. Obviously, great care must be taken when entering such a zone in order to avoid differences in potential between inside and outside.

The areas mentioned in this and the previous method are very uncommon. Where they do exist, they should be under constant supervision to ensure that no additions or alterations can lessen the protection intended.

Electrical separation

This method relies on a supply from a safety source such as an isolating transformer to BS 3535 which has no earth connection on the secondary side. In the event of a circuit that is supplied from a source developing a live fault to an exposed conductive part, there would be no path for shock current to flow (see Figure 2).

Figure 2

IP codes

First numeral: mechanical protection

0 No protection of persons against contact with live or moving parts inside the enclosure. No protection of equipment against ingress of solid foreign bodies.
1 Protection against accidental or inadvertent contact with live or moving parts inside the enclosure by a large surface of the human body, for example a hand, but not protection against deliberate access to such parts.
2 Protection against ingress of large solid foreign bodies. Protection against contact with live or moving parts inside the enclosure by fingers. Protection against ingress of medium-size solid foreign bodies.
3 Protection against contact with live or moving parts inside the enclosures by tools, wires or such objects of thickness greater than 2.5 mm. Protection against ingress of small foreign bodies.
4 Protection against contact with live or moving parts inside the enclosure by tools, wires or such objects of thickness greater than 1 mm. Protection against ingress of small solid foreign bodies.
5 Complete protection against contact with live or moving parts inside the enclosure. Protection against harmful deposits of dust. The ingress of dust is not totally prevented, but dust cannot enter in an amount sufficient to interfere with satisfactory operation of the equipment enclosed.
6 Complete protection against contact with live or moving parts inside the enclosures. Protection against ingress of dust.

Second numeral: liquid protection

0 No protection.
1 Protection against drops of condensed water. Drops of condensed water falling on the enclosure shall have no harmful effect.
2 Protection against drops of liquid. Drops of falling liquid shall have no harmful effect when the enclosure is tilted at any angle up to 15° from the vertical.
3 Protection against rain. Water falling in rain at an angle equal to or smaller than 60° with respect to the vertical shall have no harmful effect.
4 Protection against splashing. Liquid splashed from any direction shall have no harmful effect.
5 Protection against water jets. Water projected by a nozzle from any direction under stated conditions shall have no harmful effect.
6 Protection against conditions on ships' decks (deck with watertight equipment). Water from heavy seas shall not enter the enclosures under prescribed conditions.
7 Protection against immersion in water. It must not be possible for water to enter the enclosure under stated conditions of pressure and time.
8 Protection against indefinite immersion in water under specified pressure. It must not be possible for water to enter the enclosure.
X Indicates no *specified* protection.

Figure 3

Once again, great care must be taken to maintain the integrity of this type of system, as an inadvertent connection to earth, or interconnection with other circuits, would render the protection useless.

As with direct contact, RCDs are a useful back-up to EEBADS, and their use is essential when hand-held equipment is used outside the main equipotential zone.

The use of enclosures is not limited to protection against direct contact they clearly provide protection against the ingress of foreign bodies and moisture. In order to establish to what degree an enclosure can resist such ingress, reference to the IP code (BSEN 60529) should be made. Figure 3 illustrates part of the IP code.

The most commonly quoted IP codes in the 16th edition are IP2X, IPXXB and IP4X. The X denotes that protection is not specified, not that there is no protection. For example, an enclosure that was to be immersed in water would be classified IPX8, there would be no point using the code IP68.

Note: IPXXB denotes protection against finger contact only.

1
AN OVERVIEW

So, here you are outside the premises, armed with lots of test instruments, a clipboard, a pad of test results sheets, the IEE Regulations, Guidance Notes 3, and an instruction to carry out an inspection and test of the electrical installation therein. Dead easy, you've been told, piece of cake, just poke about a bit, 'Megger' the wiring, write the results down, sign the test certificate and you should be on to the next job within the hour!

Oh! would that it were that simple! What if lethal defects were missed by just 'poking about'? What if other tests should have been carried out which may have revealed serious problems? What if things go wrong after you have signed to say all is in accordance with the Regulations? What if you were not actually competent to carry out the inspection and test in the first place? What if . . . and so on, the list is endless. Inspection, testing and certification is a serious and, in many instances, a complex matter, so let us wind the clock back to the point at which you were about to enter the premises to carry out your tests, and consider the implications of carrying out an inspection and test of an installation.

What are the legal requirements in all of this? Where do you stand if things go wrong? What do you need to do to ensure compliance with the law?

Statutory regulations

The IEE Wiring Regulations (BS 7671) and associated guidance notes are *not* statutory documents, they can however be used in a court of law to prove compliance with statutory requirements such as the Electricity at Work Regulations (EAWR) 1989, which cover all work activity associated with electrical systems in non-domestic situations. A list of other statutory Regulations are given in Appendix 2 of the IEE Regulations. However, it is the Electricity at Work Regulations that are most closely associated with BS7671, and as such it is worth giving some areas a closer look.

There are thirty-three Regulations in all, twelve of which deal with the special requirements of mines and quarries, and another four with exemptions and extensions outside the UK etc. We are only concerned with the first sixteen Regulations, and Regulation 29, the defence regulation, which we shall come back to later. Let us start then, with a comment on the meaning of *electrical systems and equipment*.

Electrical systems and equipment

According to the EAWR, electrical systems and equipment can encompass anything from power stations to torch or wrist-watch batteries. A battery may not create a shock risk, but may cause burns or injury as a result of attempting to destroy it by fire, whereby explosions may occur. A system can actually include the source of energy, so, a test instrument with its own supply e.g. a continuity tester is a system in itself, and a loop impedance tester, which requires an external supply source, becomes part of the system into which it is connected. From the preceding comments it will be obvious then, that in broad terms, if something is electrical, it is or is part of, an electrical system. So, where does responsibility lie for any involvement with such a system?

The EAWR requires that every employer, employee and self-employed person be responsible for compliance with the Regulations with regards to matters within their control, and as such are known *duty holders*. Where then do you stand as the person about to conduct an inspection and test of an installation? Most certainly, you are a duty holder in that you have control of the installation in so far as you will ultimately pass the installation as safe or make recommendations to ensure its safety. You also have control of the test instruments which, as already stated are systems in themselves, and control of the installation whilst testing is being carried out.

Any breach of the Regulations may result in prosecution, and unlike the other laws, under this Act you are presumed guilty and have to establish your innocence by invoking the Defence Regulation 29. Perhaps some explanation is needed here. Each of the sixteen Regulations has a status, in that it is either *absolute* or *reasonably practicable*.

Regulations that are *absolute* must be conformed to at all cost, whereas those that are *reasonably practicable* are conformed to, provided that all reasonable steps have been taken to ensure safety. For the contravention of an *absolute* requirement, Regulation 29 is available as a defence in the event of criminal prosecution, provided the accused can demonstrate that they took all reasonable and diligent steps to prevent danger or injury.

No one wants to end up in court accused of negligence, and so we need to be sure that we know what we are doing when we are inspecting and testing.

Apart from the knowledge required competently to carry out the verification process, the person conducting the inspection and test, must be in possession of test instruments appropriate to the duty required of them.

Instruments

In order to fulfil the basic requirements for testing to BS 7671, the following instruments are needed:

1 A low-resistance ohm-meter (continuity tester).
2 An insulation resistance tester.
3 A loop impedance tester.
4 An RCD tester.
5 A prospective short circuit current (PSCC) tester.
6 An approved test lamp or voltage indicator.
7 A proving unit.

An overview

Many instrument manufacturers have developed dual or multi-function instruments, hence it is quite common to have continuity and insulation resistance in one unit, loop impedance and PSCC in one unit, loop impedance, PSCC and RCD tests in one unit etc. However, regardless of the various combinations, let us take a closer look at the individual test instrument requirements.

Low resistance ohmmeters/continuity testers

Bells, buzzers, simple multimeters etc. will all indicate whether or not a circuit is continuous, but will not show the difference between the resistance of say, a 10 m length of $10 \, mm^2$ conductor and a 10 m length of $1 \, mm^2$ conductor. I use this example as an illustration, as it is based on a real experience of testing the continuity of a $10 \, mm^2$ main bonding conductor between gas and water services. The services, some 10 m apart, were at either ends of a domestic premises. The $10 \, mm^2$ conductor, connected to both services, disappeared under the floor, and a measurement between both ends indicated a resistance higher than expected. Further investigation revealed, that just under the floor at each end, the $10 \, mm^2$ conductor had been terminated in a connector block and the join between the two, about 8 m, had been wired with a $1 \, mm^2$ conductor. Only a milli-ohmmeter would have detected such a fault.

A low resistance ohm-meter should have a no-load source voltage of between 4 V and 24 V, and be capable of delivering an AC or DC short circuit voltage of not less than 200 mA. It should have a resolution (i.e. a detectable difference in resistance) of at least 0.05 milli-ohms.

Insulation resistance testers

An *insulation resistance test* is the correct term for this form of testing, not a '*megger*' test as megger is a manufacturer's trade name, not the name of the test.

An insulation resistance tester must be capable of delivering 1 mA when the required test voltage is applied across the minimum acceptable value of insulation resistance.

Hence, an instrument selected for use on a low voltage (50 VAC–1000 VAC) system should be capable of delivering 1 mA at 500 V across a resistance of 0.5 megohms.

Loop impedance tester

This instrument functions by creating, in effect, an earth fault for a brief moment, and is connected to the circuit via a plug or by 'flying leads' connected separately to phase, neutral, and earth.

The instrument should only allow an earth fault to exist for a maximum of 40 ms, and a resolution of 0.01 ohms is adequate for circuits up to 50 A. Above this circuit rating, the ohmic values become too small to give such accuracy using a standard instrument, and more specialized equipment may be required.

RCD tester

Usually connected by the use of a plug, although 'flying leads' are needed for non-socket outlet circuits, this instrument allows a range of out of balance currents to flow through the RCD to cause its operation within specified time limits.

The test instrument should not be operated for longer than 2 seconds, and it should have a 10 per cent accuracy across the full range of test currents.

PSCC tester

Normally one half of a dual, loop impedance/PSCC tester, this instrument measures the prospective phase neutral fault current at the point of measurement using the same leads as for loop impedance.

Approved test lamp or voltage indicator

A flexible cord with a lamp attached is not an approved device, nor for that matter is the ubiquitous 'testascope' or 'neon screwdriver', which encourages the passage of current, at low voltage, through the body!

A typical approved test lamp is as shown in Figure 1.1.

The Health and Safety Executive, Guidance Note 38, recommend that the leads and probes associated with test lamps, voltage indicators, voltmeters etc. have the following characteristics:

1 The leads should be adequately insulated and, ideally, fused.
2 The leads should be easily distinguished from each other by colour.
3 The leads should be flexible and sufficiently long for their purpose.
4 The probes should incorporate finger barriers, to prevent accidental contact with live parts.
5 The probes should be insulated and have a maximum of 2 mm of exposed metal, but preferably have spring loaded enclosed tips.

Proving unit

This is an optional item of test equipment, in that test lamps should be proved on a known supply which could, of course, be an

Lamp

Maximum
test
voltage
marked

Insulated lead

Fused test
probes

Finger guards

2 mm exposed
or spring loaded
enclosed tips

Figure 1.1

adjacent socket or lighting point etc. However, to prove a test lamp on such a known supply may involve entry into enclosures with the associated hazards that such entry could bring. A proving unit is a compact device not much larger than a cigarette packet, which is capable of electronically developing 240 V DC across which the test lamp may be proved. The exception to this are test lamps incorporating 240 V lamps which will not activate from the small power source of the proving unit.

Test lamps must be proved against a voltage similar to that to be tested. Hence, proving test lamps that incorporate an internal check i.e. shorting out the probes to make a buzzer sound, is not acceptable if the voltage to be tested is higher than that delivered by the test lamp.

Care of test instruments

The Electricity at Work Regulations require that all electrical systems, this includes test instruments, be maintained to prevent danger. This does not restrict such maintenance to just a yearly calibration, but requires equipment to be kept in good condition in order that it is safe to use at all times.

Whilst test instruments and associated leads probes and clips etc., used in the electrical contracting industry are robust in design and manufacture, they still need treating with care and protecting from mechanical damage. Keep test gear in a separate box or case away from tools and sharp objects and always check the general condition of a tester and leads before they are used.

2
INITIAL INSPECTION

Inspection and testing

Circumstances which require an initial verification
New installations or Additions or Alterations

General reasons for initial verification

1 To ensure equipment and accessories are to a relevant standard.
2 To prove compliance with BS 7671
3 To ensure that the installation is not damaged so as to impair safety.

Information required
Assessment of general characteristics sections 311, 312 and 313 together with information such as drawings, charts etc. in accordance with Reg. 514–09–01.

Documentation required and to be completed
Electrical Installation Certificate signed or authenticated for the design and construction (could be the same person) and then for the inspection and test. A schedule of test results must accompany an Electrical Installation Certificate.

Sequence of tests

1 Continuity of all protective conductors
2 Continuity of ring final circuit conductors
3 Insulation resistance
4 Site applied insulation
5 Protection by separation of circuits
6 Protection against direct contact by barriers and enclosures provided during erection
7 Insulation of non-conducting floors and walls
8 Polarity
9 Earth electrode resistance
10 Earth fault loop impedance
11 Functional testing

Before any testing is carried out, a detailed physical inspection must be made to ensure that all equipment is to a relevant British or Harmonized European Standard, and that it is erected/installed in compliance with the IEE Regulations, and that it is not damaged such that it could cause danger. In order to comply with these requirements, the Regulations give a check list of some eighteen items that, where relevant, should be inspected.

However, before such an inspection, and test for that matter, is carried out, certain information *must* be available to the verifier. This information is the result of the Assessment of General Characteristics required by IEE Regulations Part 3, Sections 311, 312 and 313, and drawings, charts, and similar information relating to the installation. It is at this point that most readers who work in the real world of electrical installation will be lying on the floor laughing hysterically.

Let us assume that the designer and installer of the installation are competent professionals, and all of the required documentation is available.

Interestingly, one of the items on the check list *is* the presence of diagrams, instructions and similar information. If these are missing then there is a departure from the Regulations.

Another item on the list is the verification of conductors for current carrying capacity and voltage drop in accordance with the design. How on earth can this be verified without all the information? A 30 A Type B circuit breaker (CB) protecting a length of 4mm^2 conductor may look reasonable, but is it correct, and are you prepared to sign to say that it is unless you are sure? Let us look then at the general content of the check list.

1 **Connection of conductors** Are terminations electrically and mechanically sound, is insulation and sheathing removed only to a minimum to allow satisfactory termination?

2 **Identification of conductors** Are conductors correctly identified in accordance with the Regulations?

3 **Routing of cables** Are cables installed such that account is taken of external influences such as mechanical damage, corrosion, heat etc?

4 **Conductor selection** Are conductors selected for current carrying capacity and voltage drop in accordance with the design?

5 **Connection of single pole devices** Are single pole protective and switching devices connected in the phase conductor only?

6 **Accessories and equipment** Are all accessories and items of equipment correctly connected?

7 **Thermal effects** Are fire barriers present where required and protection against thermal effects provided?

8 **Protection against shock** What methods have been used to provide protection against direct and indirect contact?

9 **Mutual detrimental influence** Are wiring systems installed such that they can have no harmful effect on non-electrical systems, or that systems of different currents or voltages are segregated where necessary?

10 **Isolation and switching** Are there appropriate devices for isolation and switching correctly located and installed?

11 **Undervoltage** Where undervoltage may give rise for concern, are there protective devices present?

12 **Protective devices** Are protective and monitoring devices correctly chosen and set to ensure protection against indirect contact and/or overcurrent?

13 **Labelling** Are all protective devices, switches (where necessary) and terminals correctly labelled?

14 **External influences** Have all items of equipment and protective measures been selected in accordance with the appropriate external influences?
15 **Access** Are all means of access to switchgear and equipment adequate?
16 **Notices and signs** Are danger notices and warning signs present?
17 **Diagrams** Are diagrams, instructions and similar information relating to the installation available?
18 **Erection methods** Have all wiring systems, accessories and equipment been selected and installed in accordance with the requirements of the Regulations, and are fixings for equipment adequate for the environment?

So, we have now inspected all relevant items, and provided that there are no defects that may lead to a dangerous situation when testing, we can now start the actual testing procedure.

3
TESTING CONTINUITY OF PROTECTIVE CONDUCTORS

All protective conductors, including main equipotential and supplementary bonding conductors must be tested for continuity using a low resistance ohmmeter.

For main equipotential bonding there is no single fixed value of resistance above which the conductor would be deemed unsuitable. Each measured value, if indeed it is measurable for very short lengths, should be compared with the relevant value for a particular conductor length and size. Such values are shown in Table 3.1.

Table 3.1 Resistance (Ω) of copper conductors at 20°C

CSA (mm^2)	Length (m)									
	5	10	15	20	25	30	35	40	45	50
1	0.09	0.18	0.27	0.36	0.45	0.54	0.63	0.72	0.82	0.9
1.5	0.06	0.12	0.18	0.24	0.3	0.36	0.43	0.48	0.55	0.6
2.5	0.04	0.07	0.11	0.15	0.19	0.22	0.26	0.03	0.33	0.37
4	0.023	0.05	0.07	0.09	0.12	0.14	0.16	0.18	0.21	0.23
6	0.02	0.03	0.05	0.06	0.08	0.09	0.11	0.13	0.14	0.16
10	0.01	0.02	0.03	0.04	0.05	0.06	0.063	0.07	0.08	0.09
16	0.006	0.01	0.02	0.023	0.03	0.034	0.04	0.05	0.05	0.06
25	0.004	0.007	0.01	0.015	0.02	0.022	0.026	0.03	0.033	0.04
35	0.003	0.005	0.008	0.01	0.013	0.016	0.019	0.02	0.024	0.03

Where a supplementary equipotential bonding conductor has been installed between *simultaneously accessible* exposed and extraneous conductive parts, because circuit disconnection times cannot be met, then the resistance (R) of the conductor, must be equal to or less than 50/Ia. In the case of construction sites and agricultural or horticultural installations the 50 is replaced by 25.

So, $R \leqslant$ 50/Ia, or 25/Ia, where 50, and 25, are the voltages above which exposed metalwork should not rise, and Ia is the minimum current causing operation of the circuit protective device within 5 secs.

For example, suppose a 45 A BS 3036 fuse protects a cooker circuit, the disconnection time for the circuit cannot be met, and so a supplementary bonding conductor has been installed between the cooker case and the adjacent metal sink. The resistance (R) of that conductor should not be greater than 50/Ia, and Ia in this case is 145 A (see Figure 3.2B of the IEE Regulations)

i.e. $50/145 = 0.34 \, \Omega$

How then, do we conduct a test to establish continuity of main or supplementary bonding conductors? Quite simple really, just connect the leads from a low resistance ohmmeter to the ends of the bonding conductor (Figure 3.1). One end should be disconnected from its bonding clamp, otherwise any measurement may include

Figure 3.1

14

the resistance of parallel paths of other earthed metalwork. Remember to zero the instrument first or, if this facility is not available, record the resistance of the test leads so that this value can be subtracted from the test reading.

IMPORTANT NOTE: If the installation is in operation, then *never* disconnect main bonding conductors unless the supply can be isolated. Without isolation, persons and livestock are at risk of electric shock.

The continuity of circuit protective conductors may be established in the same way, but a second method is preferred, as the results of this second test indicate the value of $(R_1 + R_2)$ for the circuit in question.

The test is conducted in the following manner:

1 Temporarily link together the phase conductor and cpc of the circuit concerned in the distribution board or consumer unit.
2 Test between phase and cpc at each outlet in the circuit. A reading indicates continuity.
3 Record the test result obtained at the furthest point in the circuit. This value is $(R_1 + R_2)$ for the circuit.

Figure 3.2 illustrates the above method.

Figure 3.2

Reading at last outlet gives the value of (R1 + R2) for the circuit

There may be some difficulty in determining the $(R_1 + R_2)$ values of circuits in installations that comprise steel conduit and trunking, and/or SWA and mims cables because of the parallel earth paths that are likely to exist. In these cases, continuity tests may have to be carried out at the installation stage before accessories are connected or terminations made off as well as after completion.

Although it is no longer considered good working practice to use steel conduit or trunking as a protective conductor, it is permitted, and hence its continuity must be proved. The enclosure must be inspected along its length to ensure that it is sound and then the standard low resistance test is performed. If the verifier has any doubt as to the soundness of the conductor, a further test is made using a high current test instrument which has a test voltage not exceeding 50 V and can deliver up to 1.5 times the design current of the circuit up to a maximum of 25 A. This test can cause arcing at faulty joints and hence should not be carried out if there is any chance of danger.

4
TESTING CONTINUITY OF RING FINAL CIRCUIT CONDUCTORS

There are two main reasons for conducting this test:

1 To establish that interconnections in the ring do not exist.
2 To ensure that the cpc is continuous, and indicate the value of $(R_1 + R_2)$ for the ring.

What then are interconnections in a ring circuit, and why is it important to locate them? Figure 4.1 shows a ring final circuit with an interconnection.

The most likely cause of the situation shown in Figure 4.1 is where a DIY enthusiast has added sockets P, Q, R and S to an existing ring A, B, C, D, E and F.

Figure 4.1

In itself there is nothing wrong with this. The problem arises if a break occurs at, say, point Y, or the terminations fail in socket C or P. Then there would be four sockets all fed from the point X which would then become a spur.

So, how do we identify such a situation with or without breaks at point 'Y'? A simple resistance test between the ends of the phase, neutral or circuit protective conductors will only indicate that a circuit exists, whether there are interconnections or not. The following test method is based on the philosophy that the resistance measured across any diameter of a perfect circle of conductor will always be the same value (Figure 4.2).

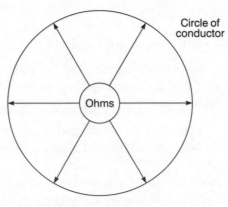

Same value whatever diameter is measured

Figure 4.2

The perfect circle of conductor is achieved by cross connecting the phase and neutral legs of the ring (Figure 4.3).

The test procedure is as follows:

1 Identify the opposite legs of the ring. This is quite easy with sheathed cables, but with singles, each conductor will have to be identified, probably by taking resistance measurements between each one and the closest socket outlet. This will give three high readings and three low readings thus establishing the opposite legs.

Figure 4.3

2 Take a resistance measurement between the ends of each conductor loop. Record this value.

3 Cross connect the opposite ends of the phase and neutral loops (Figure 4.4).

4 Measure between phase nd neutral at each socket on the ring. The readings obtained should be, for a perfect ring, substantially the same. If an interconnection existed such as shown in Figure 4.1, then sockets A to F would all have similar readings, and those beyond the interconnection would have gradually increasing values to approximately the mid point of the ring, then decreasing

Figure 4.4

values back towards the interconnection. If a break had occurred at point Y then the readings from socket S would increase to a maximum at socket P. One or two high readings are likely to indicate either loose connections or spurs. A null reading, i.e. an open circuit indication, is probably a reverse polarity, either phase-cpc or neutral-cpc reversal. These faults would clearly be rectified and the test at the suspect socket(s) repeated.

5 Repeat the above procedure, but in this case cross connect the phase and cpc loops (Figure 4.5).

Figure 4.5

In this instance, if the cable is of the flat twin type, the readings at each socket will increase very slightly and then decrease around the ring. This difference, due to the phase and cpc being different sizes, will not be significant enough to cause any concern. The measured value is very important, it is $R_1 + R_2$ for the ring.

As before, loose connections, spurs and, in this case, P – N cross polarity, will be picked up.

Table 4.1

Initial measurements	P1–P2	N1–N2	cpc 1 – cpc2
	0.52	0.52	0.86
Reading at each socket	0.26	0.26	0.32 – 0.34
For spurs, each metre in length will add the following resistance to the above values	0.015	0.015	0.02

Socket	P–N Reading	P–CPC Reading
A	OK	No Reading
B	No Reading	OK
C	No Reading	No Reading

A — P – N reversal or twisted CPC's not in terminal

B — P – CPC reversal or twisted N's not in terminal

C — N – CPC reversal or twisted P's not in terminal

Figure 4.6

The details in Table 4.1 are typical approximate ohmic values for a healthy 70 m ring final circuit wired in 2.5/1.5 flat twin and cpc cable. (In this case the CPC will be approximately 1.67× the P or N resistance.)

As already mentioned null readings may indicate a reverse polarity. They could also indicate twisted conductors not in their terminal housing. The examples shown in Figure 4.6 may help to explain these situations.

5
TESTING INSULATION RESISTANCE

This is probably the most used and yet abused test of them all. Affectionately known as 'meggering', an *insulation resistance test* is performed in order to ensure that the insulation of conductors, accessories and equipment is in a healthy condition, and will prevent dangerous leakage currents between conductors and between conductors and earth. It also indicates whether any short circuits exist.

Insulation resistance, as just discussed, is the resistance measured between conductors and is made up of countless millions of resistances in parallel (Figure 5.1).

The more resistances there are in parallel, the *lower* the overall resistance, and in consequence, the longer a cable the lower the insulation resistance. Add to this the fact that almost all installation circuits are also wired in parallel, it becomes apparent that tests on large installations may give, if measured as a whole, pessimistically low values, even if there are no faults.

Under these circumstances, it is usual to break down such large installations into smaller sections, floor by floor, sub-main by sub-main etc. This also helps, in the case of periodic testing, to minimize disruption. The test procedure is as follows:

Conductors

Insulation

Resistance of
insulation in parallel

Figure 5.1

1 Disconnect all items of equipment such as capacitors and indicator lamps as these are likely to give misleading results. Remove any items of equipment likely to be damaged by the test, such as dimmer switches, electronic timers etc. Remove all lamps and accessories and disconnect fluorescent and discharge fittings. Ensure that the installation is disconnected from the supply, all fuses are in place, and MCBs and switches are in the on position. In some instances it may be impracticable to remove lamps etc. and in this case the local switch controlling such equipment may be left in the off position.

2 Join together all live conductors of the supply and test between this join and earth. Alternatively, test between each live conductor and earth in turn.

3 Test between phase and neutral. For three phase systems, join together all phases and test between this join and neutral. Then test between each of the phases. Alternatively, test between each of the live conductors in turn. Installations incorporating two-way lighting systems should be tested twice with the two-way switches in alternative positions.

23

Table 5.1

System	Test voltage	Minimum insulation resistance
SELV and PELV	250 V DC	0.25 MΩ
LV up to 500 V	500 V DC	0.5 MΩ
Over 500 V	1000 V DC	1.0 MΩ

Table 5.1 gives the test voltages and minimum values of insulation resistance for ELV and LV systems.

If a value of less than $2\,M\Omega$ is recorded it may indicate a situation where a fault is developing, but as yet still complies with the minimum permissible value. In this case each circuit should be tested separately and each should be above $2\,M\Omega$.

Example

An installation comprising six circuits have individual insulation resistances of $2.5\,M\Omega$, $8\,M\Omega$, $200\,M\Omega$, $200\,M\Omega$, $200\,M\Omega$ and $200\,M\Omega$, and so the total insulation resistance will be:

$$\frac{1}{R_t} = \frac{1}{2.5} + \frac{1}{8} + \frac{1}{200} + \frac{1}{200} + \frac{1}{200} + \frac{1}{200}$$

$$= 0.4 + 0.125 + 0.005 + 0.005 + 0.005 + 0.005$$

$$= 0.545$$

$$R_t = \frac{1}{0.545}$$

$$= 1.83\,M\Omega$$

This is clearly greater than the $0.5\,M\Omega$ minimum but less than $2\,M\Omega$, but as all circuits are greater than $2\,M\Omega$ the system could be considered satisfactory.

6
SPECIAL TESTS

The next four tests are special in that they are not often required in the general type of installation. They also require special test equipment. In consequence, the requirements for these tests will only be briefly outlined in this chapter.

Site applied insulation

When insulation is applied to live parts during the erection process on site in order to provide protection against direct contact, then a test has to be performed to show that the insulation can withstand a high voltage equivalent to that specified in the BS for similar factory built equipment.

If supplementary insulation is applied to equipment on site, to provide protection against indirect contact, then the voltage withstand test must be applied, and the insulating enclosure must afford a degree of protection of not less than IP2X or IPXXB.

Protection by separation of circuits

When SELV or PELV is used as a protective measure, then the separation from circuits of a higher voltage has to be verified by an insulation resistance test at a test voltage of 250 V and result in a minimum insulation resistance of 0.25 MΩ. If the circuit is at low voltage and supplied from, say, a BS 3535 transformer the test is at 500 V with a minimum value of 0.555 MΩ.

Protection by barriers or enclosures

If, on site, protection against direct contact is provided by fabricating an enclosure or erecting a barrier, it must be shown that the enclosure can provide a degree of protection of at least IP2X or IPXXB. Readily accessible horizontal top surfaces should be to at least IP4X.

An enclosure having a degree of protection IP2X can withstand the ingress of fingers and solid objects exceeding 12 mm diameter. IPXXB is protection against finger contact only. IP4X gives protection against wires and solid objects exceeding 1 mm in diameter.

The test for IP2X or IPXXB is conducted with a 'standard test finger' which is supplied at a test voltage not less than 40 V and no more than 50 V. One end of the finger is connected in series with a lamp and live parts in the enclosure. When the end of the finger is introduced into the enclosure, provided the lamp does not light then the protection is satisfactory.

The test for IP4X is conducted with a rigid 1 mm diameter wire with its end bent at right angles. Protection is afforded if the wire does not enter the enclosure.

Protection by non-conducting location

This is a rare location and demands specialist equipment to measure the insulation resistance between insulated floors and walls at various points.

7
TESTING POLARITY

This simple test, often overlooked, is just as important as all the others, and many serious injuries and electrocutions could have been prevented if only polarity checks had been carried out.

The requirements are:

1 All fuses and single pole switches are in the phase conductor.
2 The centre contact of an Edison screw type lampholder is connected to the phase conductor.
3 All socket outlets and similar accessories are correctly wired.

Although polarity is towards the end of the recommended test sequence, it would seem sensible, on lighting circuits, for example, to conduct this test at the same time as that for continuity of CPCs (Figure 7.1).

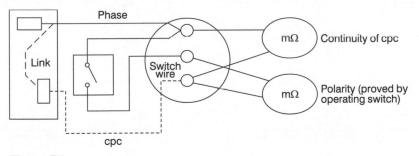

Figure 7.1

Testing polarity

As discussed earlier, polarity on ring final circuit conductors, is achieved simply by conducting the ring circuit test. For radial socket outlet circuits, however, this is a little more difficult. The continuity of the CPC will have already been proved by linking phase and CPC and measuring between the same terminals at each socket. Whilst a phase-CPC reversal would not have shown, a phase-neutral reversal would, as there would have been no reading at the socket in question. This would have been remedied, and so only phase-CPC reversals need to be checked. This can be done by linking together phase and neutral at the origin and testing between the same terminals at each socket. A phase-CPC reversal will result in no reading at the socket in question.

When the supply is connected, it is important to check that the incoming supply is correct. This is done using an approved voltage indicator at the intake position or close to it.

8
TESTING EARTH FAULT LOOP IMPEDANCE

This is very important but sadly, poorly understood. So let us remind ourselves of the component parts of the earth fault loop path (Figure 8.1). Starting at the point of fault:

1 The CPC.
2 The main earthing conductor and earthing terminal.
3 The return path via the earth for TT systems, and the metallic return path in the case of TN–S or TN–C–S systems. In the latter case the metallic return is the PEN conductor.
4 The earthed neutral of the supply transformer.
5 The transformer winding.
6 The phase conductor back to the point of fault.

Overcurrent protective devices must, under earth fault conditions, disconnect fast enough to reduce the risk of electric shock. This is achieved if the actual value of the earth fault loop impedance does not exceed the tabulated maximum values given in the IEE regulations.

The purpose of the test, therefore, is to determine the actual value of the loop impedance (Z_s), for comparison with those maximum values, and it is conducted as follows:

1 Ensure that all main equipotential bonding is in place.
2 Connect the test instrument either by its BS 4363 plug, or the 'flying leads', to the phase, neutral and earth terminals at

Figure 8.1

the remote end of the circuit under test. (If a neutral is not available, e.g. in the case of a three phase motor, connect the neutral probe to earth.)

3 Press to test and record the value indicated.

It must be understood, that this instrument reading is *not valid for direct comparison with the tabulated maximum values*, as account must be taken of the ambient temperature at the time of test, and the maximum conductor operating temperature, both of which will have an effect on conductor resistance. Hence, the $(R_1 + R_2)$ could be greater at the time of fault than at the time of test.

So, our measured value of Z_s must be corrected to allow for these possible increases in temperature occurring at a later date. This requires actually measuring the ambient temperature and applying factors in a formula.

Clearly this method of correcting Z_s is time consuming and unlikely to be commonly used. Hence, a rule of thumb method may be applied which simply requires that the measured value of Z_s does not exceed 3/4 of the appropriate tabulated value. Table 8.3 gives the 3/4 values of tabulated loop impedance for direct comparison with measured values.

Table 8.1 Values of loop impedance for comparison with test readings

Protection	Disconnection time (s)		5 A	6 A	10 A	15 A	16 A	20 A	25 A	30 A	32 A	40 A	45 A	50 A	60 A	63 A	80 A	100 A	125 A	160 A	200 A
BS 3036 fuse	0.4	Z_s max	7.5	–	–	2	–	1.38	–	0.85	–	–	0.46								
	5	Z_s max	13.9	–	–	4.1	–	3	–	2.97	–	–	1.25	–	0.87			0.42			
BS 88 fuse	0.4	Z_s max	–	6.66	4	–	2.11	1.38	1.12	–	0.82	0.64	–	0.47							
	5	Z_s max	–	10.5	5.8	–	3.27	2.28	1.8	–	1.44	1.05	–	0.82	–	0.64	0.45	0.33	0.26	0.2	0.14
BS 1361 fuse	0.4	Z_s max	8.17	–	–	2.57	–	1.33	–	0.9	–	–	0.45								
	5	Z_s max	12.8	–	–	3.9	–	2.19	–	1.44	–	–	0.75	0.54	–	0.39	0.28				
BS 3871 MCB Type 1	0.4 & 5	Z_s max	9	7.5	4.5	3	2.81	2.25	1.8	1.5	1.41	1.12	1	0.9	–	0.71					
BS 3871 MCB Type 2	0.4 & 5	Z_s max	5.14	4.28	2.57	1.71	1.6	1.28	1.02	0.85	0.8	0.64	0.57	0.51	–	0.4					
BS 3871 MCB Type 3	0.4 & 5	Z_s max	3.6	3	1.8	1.2	1.12	0.9	0.72	0.6	0.56	0.45	0.4	0.36	–	0.28					
BS EN 60898 CB Type B	0.4 & 5	Z_s max	–	6	3.6	–	2.25	1.8	1.44	–	1.12	0.9	0.8	0.72	–	0.57					
BS EN 60898 CB Type C	0.4 & 5	Z_s max	3.6	3	1.8	1.2	1.12	0.9	0.72	0.6	0.56	0.45	0.4	0.36	–	0.28					
BS EN 60898 CB Type D	0.4 & 5	Z_s max	1.8	1.5	0.9	0.6	0.56	0.45	0.36	0.3	0.28	0.22	0.2	0.18	–	0.14					

Rating of protection

In effect, a loop impedance test places a phase/earth fault on the installation, and if an RCD is present it may not be possible to conduct the test as the device will trip out each time the loop impedance tester button is pressed. Unless the instrument is of a type that has a built-in guard against such tripping, the value of Z_s will have to be determined from measured values of Z_e and $(R_1 + R_2)$, and the 3/4 rule applied.

IMPORTANT NOTE: Never short out an RCD in order to conduct this test.

As a loop impedance test creates a high earth fault current, albeit for a short space of time, some lower rated MCBs may operate resulting in the same situation as with an RCD, and Z_s will have to be calculated. It is not really good practice temporarily to replace the MCB with one of a higher rating.

External loop impedance Z_e

The value of Z_e is measured at the intake position on the supply side and with all main equipotential bonding disconnected. Unless the installation can be isolated from the supply, this test should not be carried out, as a potential shock risk will exist with the supply on and the main bonding disconnected.

9
TESTING EARTH
ELECTRODE
RESISTANCE

In many rural areas, the supply system is TT and hence reliance is placed on the general mass of earth for a return path under earth fault conditions. Connection to earth is made by an electrode, usually of the rod type, and preferably installed as shown in Figure 9.1.

In order to determine the resistance of the earth return path, it is necessary to measure the resistance that the electrode has with earth. If we were to make such measurements at increasingly longer distances from the electrode, we would notice an increase in

Figure 9.1

33

Figure 9.2

resistance up to about 2.5–3 m from the rod, after which no further increase in resistance would be noticed (Figure 9.2).

The maximum resistance recorded is the electrode resistance and the area that extends the 2.5–3 m beyond the electrode is known as the earth electrode resistance area.

There are two methods of making the measurement, one using a proprietary instrument, and the other using a loop impedance tester.

Method 1 – protection by overcurrent device

This method is based on the principle of the potential divider (Figure 9.3).

By varying the position of the slider the resistance at any point may be calculated from $R = V/I$.

34

Figure 9.3

The earth electrode resistance test is conducted in a similar fashion, with the earth replacing the resistance and a potential electrode replacing the slider (Figure 9.4). In Figure 9.4 the earthing conductor to the electrode under test is temporarily disconnected.

The method of test is as follows:

1 Place the current electrode (C2) away from the electrode under test, approximately 10 times its length, i.e. 30 m for a 3 m rod.
2 Place the potential electrode mid way.
3 Connect test instrument as shown.
4 Record resistance value.

Figure 9.4

5 Move the potential electrode approximately 3 m either side of the mid position, and record these two readings.
6 Take an average of these three readings (this is the earth electrode resistance).
7 Determine the maximum deviation or difference of this average from the three readings.
8 Express this deviation as a percentage of the average reading.
9 Multiply this percentage deviation by 1.2.
10 Provided this value does not exceed a figure of 5% then the accuracy of the measurement is considered acceptable.

If three readings obtained from an earth electrode resistance test were 181 Ω, 185 Ω and 179 Ω. What is the value of the electrode resistance and is the accuracy of the measurement acceptable?

$$\text{Average value} = \frac{181 + 185 + 179}{3}$$

$$= 181.67 \, \Omega$$

$$\text{Maximum deviation} = 185 - 181.67$$

$$= 3.33$$

$$\text{Expressed as a percentage of the average} = \frac{3.33 \times 100}{181.67}$$

$$= 1.83\%$$

$$\text{Measurement accuracy} = 1.83\% \times 1.2 = 2.2\%$$
$$\text{(which is acceptable)}$$

For TT systems the result of this test will indicate compliance if the product of the electrode resistance and the operating current of the overcurrent device does not exceed 50 V.

Method 2 – protection by a residual current device

In this case, an earth fault loop impedance test is carried out between the incoming phase terminal and the electrode (a standard test for Z_e).

The value obtained is added to the cpc resistance of the protected circuits and this value is multiplied by the operating current of the RCD. The resulting value should not exceed 50 V. If it does, then Method 1 should be used to check the actual value of the electrode resistance.

10
FUNCTIONAL TESTING

RCD operation

Where RCDs are fitted, it is essential that they operate within set parameters. The RCD testers used are designed to do just this, and the basic tests required are as follows:

1 Set the test instrument to the rating of the RCD.
2 Set the test instrument to half rated trip.
3 Operate the instrument and the RCD should not trip.
4 Set the instrument to deliver the full rated tripping current of the RCD.
5 Operate the instrument and the RCD should trip out in the required time.

Table 10.1

RCD type	½ rated	Full trip current
BS 4239 and BS 7288 sockets	No trip	<200 ms
BS 4239 with time delay	No trip	½ time delay + 200 ms-time delay + 200 ms
BS EN 61009 or BS EN 61009 rcbo	No trip	300 ms
As above but Type S with time delay	No trip	130–500 ms

When an RCD is used for supplementary protection against direct contact, or to protect users of hand-held equipment outside the equipotential zone, it must be rated at 30 mA or less and operate within 40 ms when subjected to a tripping current of 150 mA.

Test instruments have the facility to provide this value of tripping current. There is no point in conducting this so-called 'fast trip' test if an RCD has a rating in excess of 30 mA.

All RCDs have a built-in test facility in the form of a test button. Operating this test facility creates an artificial out of balance condition that causes the device to trip. This only checks the mechanics of the tripping operation, it is not a substitute for the tests just discussed.

All other items of equipment such as switchgear, controlgear interlocks etc. must be checked to ensure that they are correctly mounted and adjusted and that they function correctly.

11
PERIODIC INSPECTION

Periodic inspection and testing

Circumstance which require a periodic inspection and test

Test and inspection is due; insurance, mortgage, licensing reasons; change of use; change of ownership; after additions or alterations; after damage; change of loading; to assess compliance with current Regulations.

General reasons for a periodic inspection and test

1 To ensure the safety of persons and livestock.
2 To ensure protection of property from fire and heat.
3 To ensure that the installation is not damaged so as to impair safety.
4 To ensure that the installation is not defective and complies with the current Regulations.

General areas of investigation

Safety; wear and tear; corrosion; damage; overloading; age; external influences; suitability; effectiveness.

Documentation to be completed

Periodic test report and schedule of test results.

Sequence of tests

1 Continuity of all protective conductors.
2 Polarity.
3 Earth fault loop impedance.
4 Insulation resistance.
5 Operation of isolating and switching devices.
6 Operation of RCDs.
7 Prospective fault current.

and where appropriate

8 Continuity of ring final circuit conductors.
9 Earth electrode resistance.
10 Manual operation of RCDs.
11 Protection by separation of circuits.
12 Insulation of non-conducting floors and walls.

This could be so simple. As it is, periodic inspection and testing tends to be complicated and frustrating. On the domestic scene, I doubt if any house owner actually decides to have a regular inspection. They say, 'If it works it must be OK'. It is usually only when there is a change of ownership that the mortgage companies insist on an electrical survey. The worst cases are, however, industry and commerce. Periodic inspections are requested, reluctantly, to satisfy insurers or an impending visit by the HSE. Even then it is usually the case that 'you can't turn that off' or 'why can't you just test this bit and then issue a certificate for the whole lot'. Under the rare circumstances when an inspection and test is genuinely requested it is difficult to convince the client that, as there are no drawings, or information about the installation, and that no switchgear is labelled etc., you are going to be on site for a considerable time and at a considerable cost.

When there are no drawings or items of information, especially on a large installation, there may be a degree of exploratory work to be carried out in order to ensure safety whilst inspecting and testing. If it is felt that it may be unsafe to continue with the inspection and test, then drawings and information **must** be produced in order to avoid contravening the Health and Safety at Work Act Section 6.

However, let us assume, as with the initial inspection, that the original installation was erected in accordance with the 16th edition, and that any alterations and/or additions have been faithfully recorded on the original documentation which is, of course, readily available!

A periodic inspection and test under these circumstances should be relatively easy, as little dismantling of the installation will be necessary, and the bulk of the work will be inspection.

Inspection should be carried out with the supply disconnected as it may be necessary to gain access to wiring in enclosures etc. and hence, with large installations it will probably need considerable liaison with the client to arrange convenient times for interruption of supplies to various parts of the installation.

This is also the case when testing protective conductors, as these must *never* be disconnected unless the supply can be isolated. This is particularly important for main equipotential bonding conductors which need to be disconnected in order to measure Z_e.

In general an inspection should reveal:

1 Any aspects of the installation that may impair the safety of persons and livestock against the effects of electric shock and burns.
2 That there are no installation defects that could give rise to heat and fire and hence damage property.
3 That the installation is not damaged or deteriorated so as to impair safety.
4 That any defects or non-compliance with the Regulations, that may give rise to danger, are identified.

As was mentioned earlier, dismantling should be kept to a minimum and hence a certain amount of sampling will take place. This sampling would need to be increased in the event of defects being found.

From the testing point of view, not all of the tests carried out on the initial inspection may need to be applied. This decision depends on the condition of the installation.

The continuity of protective conductors is clearly important as is insulation resistance and loop impedance, but one wonders if polarity tests are necessary if the installation has remained undisturbed since the last inspection. The same applies to ring circuit continuity as the P–N test is applied to detect inter-connections in the ring, which would not happen on their own.

12
CERTIFICATION

Having completed all the inspection checks and carried out all the relevant tests, it remains to document all this information. This is done on electrical installation certificates, inspection schedules, test schedules, test result schedules, periodic inspection and test reports, minor works certificates and any other documentation you wish to append to the foregoing. Examples of such documentation are shown in the IEE Guidance Notes 3 on inspection and testing.

This documentation is vitally important. It has to be correct and signed or authenticated by a competent person. Electrical installation certificates and periodic reports must be accompanied by a schedule of test results for them to be valid. It should be noted that three signatures are required on an electrical installation certificate, one in respect of the design, one in respect of the construction and one in respect of the inspection and test. (For larger installations there may be more than one designer, hence the certificate has space for two signatures, i.e. designer 1 and designer 2.) It could be, of course, that for a very small company, one person signs all three parts. Whatever the case, the original must be given to the person ordering the work, and a duplicate retained by the contractor.

One important aspect of the electrical installation certificate is the recommended interval between inspections. This should be evaluated by the designer and will depend on the type of installation and its usage. In some cases the time interval is mandatory, especially where environments are subject to use by the public. Guidance Notes 3 give recommended maximum frequencies between inspections.

A periodic report form is very similar in part to an electrical installation certificate in respect of details of the installation, i.e. maximum demand, type of earthing system, Z_e, etc. The rest of the form deals with the extent and limitations of the inspection and test, recommendations, and a summary of the installation. The record of the extent and limitations of the inspection is very important. It must be agreed with the client or other third party exactly what parts of the installation will be covered by the report and those that will not. The interval until the next test is determined by the inspector.

With regard to the schedule of test results, test values should be recorded unadjusted, any compensation for temperature etc. being made after the testing is completed.

Any alterations or additions to an installation will be subject to the issue of an electrical installation certificate, except where the addition is, say, a single point added to an existing circuit, then the work is subject to the issue of a minor works certificate.

APPENDIX A
SAMPLE PAPER

Section A short answer

1 Indicate three main areas, about which you would require information, in order correctly to carry out an initial verification of a new installation.

2 There are various documents that are relevant to the inspection and testing of an installation, state:

(a) one statutory item of documentation
(b) two non-statutory items of documentation.

3 An Electrical Installation Certificate should be accompanied by signed documentation regarding three stages of an installation. What are these stages?

4 State three areas of investigation that you would consider when carrying out a periodic inspection and test of an installation.

5 State three human senses that could be used during an inspection of an installation.

6 During a test on an installation, the following readings were obtained:

$20\,M\Omega$; $8\,kA$; $22\,ms$

List the instruments which gave these readings.

46

7 The following circuits are to be tested for insulation resistance. State the test voltages to be applied and the minimum acceptable value of insulation resistance in each case:

(a) SELV circuit
(b) LV circuit up to 500 V
(c) LV circuit over 500 V.

8 List the first three tests that should be carried out during an initial verification on a new domestic installation.

9 The test for the continuity of a cpc in a radial circuit feeding one socket outlet, uses a temporary link and a milli-ohmmeter, state:

(a) where the temporary link is connected
(b) where the milli-ohmmeter is connected
(c) what does the meter reading represent?

10 List three different protective conductors that would need to be connected to the main earthing terminal of an installation.

11 The following readings were obtained during the initial tests on a healthy ring final circuit:

P1–P2–0.8 Ω; N1–N2–0.8 Ω;
cpc1–cpc2–0.8 Ω

(a) what readings would you expect:
   (i) between P and N conductors at each socket outlet?
   (ii) between P and cpc at each socket outlet?
(b) what the P to cpc reading represents?

12 What happens to:

(a) conductor resistance when conductor length increases?
(b) insulation resistance when cable length increases?
(c) conductor resistance when conductor area increases?

13 List three precautions to be taken prior to commencing an insulation resistance test on an installation.

14 An enclosure has been fabricated on site to house electrical equipment. State the IP codes that the enclosure should at least

comply with.

15 What degree of protection is offered by enclosures offering the following:

(a) IP XXB
(b) IP4X
(c) IPX8.

16 List three reasons for conducting a polarity test on an installation.

17 What earthing systems are attributed to the following:

(a) an overhead line supply with no earth?
(b) a multicore supply cable with a separate neutral and earth?
(c) a supply cable in which the functions of earth and neutral are performed by one conductor?

18 State three locations where special considerations should be made with regards to electrical installations.

19 From the formula

$$Z_s = Z_e + \frac{(R_1 + R_2) \times 1.2 \times L}{1000}$$

what is represented by:

(a) Z_e?
(b) R_2?
(c) 1.2?

20 State any three functional tests that may be carried out on a domestic installation.

Section B

Figure A1 shows the layout of the electrical installation in a new detached garage. You are to carry out an initial verification of that installation.

1 (a) What documentation/information will you require in order to carry out the verification?
   (b) Where should it be located?
   (c) What particularly important details regarding this installation should have been included on such documentation?
   (d) What consideration should be given to the existing installation from which this new installation is fed?

2 List five areas of inspection for this installation that should be carried out prior to testing.

3 The following test results were obtained from a ring final circuit continuity test. State if the readings for each socket are

Figure A1

satisfactory and give reasons for those readings you feel are
unsatisfactory.

Phase, neutral and cpc loops = 0.5 Ω

Socket	P–N	P-cpc
A	0.25	0.26
B	No reading	0.25
C	0.35	0.24
D	0.24	No reading
E	0.26	0.26

4 (a) Describe in detail how you would carry out an insulation
resistance test on this installation.
(b) The test result indicates an overall value of 1.75 MΩ, what
actions, if any, should be taken. Explain your reasons.

5 A loop impedance test on the lighting circuit cannot be
conducted, as the 6A Type B MCB keeps tripping out. Explain
why this is, and how the problem may be overcome in order to
conduct the test.

6 (a) The electronic door sensor/motor is wired on its own radial
circuit, list all the component parts of the earth fault loop
path associated with this circuit in the event of a fault to
earth.
(b) If the maximum value of loop impedance for this circuit is
2.4 Ω and an earth fault causes a current of 120 A, show by
calculation if this value will disconnect the circuit in the
required time.

APPENDIX B
SUGGESTED SOLUTIONS
TO SAMPLE PAPER

Section A

1 Any three of:
- maximum demand
- earthing systems
- nature of supply
- diagrams, charts etc.

(3 marks)

2 (a) *The Electricity at Work Regulations*   **(1 mark)**
  (b) any two of:
- BS 7671
- *Guidance Note 3*
- *The On-Site Guide*.

(2 marks)

3
- design
- construction
- inspection and testing.

(3 marks)

4 Any three of:
- safety of persons and livestock against shock
- protection from fire and heat
- deterioration and/or damage
- no defects or non-compliance which may cause danger.

(3 marks)

Appendix B   Suggested solutions to sample paper

5 ● visual
  ● touch
  ● smell.

(3 marks)

6 ● insulation resistance tester
  ● prospective short circuit current tester
  ● RCD tester.

(3 marks)

7 ● 250 V and 0.25 MΩ
  ● 500 V and 0.5 MΩ
  ● 1000 V and 1.0 MΩ

(3 marks)

8 ● continuity of protective conductors
  ● continuity of ring final circuit conductors
  ● insulation resistance.

(3 marks)

9 ● between P and E at the consumer unit
  ● between P and E at the socket outlet
  ● this value is $(R_1 + R_2)$ for the circuit.

(3 marks)

10 Any three of:
  ● circuit protective conductor
  ● main equipotential bonding conductor
  ● earthing conductor
  ● lightning conductor.

(3 marks)

11 ● 0.4 Ω
  ● 0.4 Ω
  ● $(R_1 + R_2)$ for the ring.

(3 marks)

12 ● increases
  ● decreases
  ● decreases.

(3 marks)

13 Any three of:
  ● check on existence of electronic equipment
  ● check there are no neons, capacitors etc. in circuit

- all switches closed and accessories equipment removed
- no danger to persons or livestock by conducting the test.

(3 marks)

14 ● IP 2X
   ● IP XXB
   ● IP 4X

(3 marks)

15 ● finger contact only
   ● small foreign solid bodies or 1 mm diameter wires
   ● total submersion.

(3 marks)

16 ● all single pole devices in phase conductor only
   ● centre contact of Edison screw lampholders in phase conductor
   ● all accessories correctly connected.

(3 marks)

17 ● TT
   ● T–N–S
   ● TN–C–S

(3 marks)

18 Any three of:
   ● bathrooms and shower basins
   ● hot air saunas
   ● swimming pools
   ● construction sites
   ● agricultural and horticultural situations
   ● restrictive conductive locations
   ● high earth leakage locations
   ● caravans
   ● caravan parks
   ● highway power supplies and street furniture.

(3 marks)

19 ● external loop impedance
   ● resistance of cpc
   ● multiplier for conductor operating temperature.

(3 marks)

20 Any three of:

- test button operation of an RCD
- operation of dimmer switch
- operation of main isolating switch
- operation of MCBs
- operation of two-way switching.

(3 marks)

Section B

1 (a) The results of the assessment of general characteristics section 311, 312 and 313, and diagrams charts and similar information regarding the installation

(5 marks)

(b) In or adjacent to the distribution board   **(3 marks)**

(c) Reference to the electronic door sensor and the PIR controlled external luminaires as these could be vulnerable to a typical test   **(3 marks)**

(d) Maximum demand, rating of consumer unit, earthing and bonding arrangements, capacity of main protective device etc.   **(4 marks)**

2 Any five relevant areas from the check list.   **(3 marks each)**

Items inspected

☐ 1 Connection of conductors

☐ 2 Identification of conductors

☐ 3 Routing of cables in safe zones or protected against mechanical damage

☐ 4 Selection of conductors for current and voltage drop

☐ 5 Connection of single-pole devices for protection or switching in phase conductors only

☐ 6 Correct connection of socket-outlets and lampholders

☐ 7 Presence of fire barriers and protection against thermal effects

8 Method of protection against electric shock
  (a) Protection against both direct and indirect contact
☐   SELV
☐   Limitation of discharge of energy
  (b) Protection against direct contact
☐   Insulation of live parts
☐   Barrier or enclosure
☐   Obstacles
☐   Placing out of reach
☐   PELV
  (c) Protection against indirect contact
      (i) Earthed equipotential bonding and automatic disconnection of supply
☐   Presence of earthing conductors
☐   Presence of protective conductors
☐   Presence of main equipotential bonding conductors
☐   Presence of supplementary equipotential bonding conductors
      (ii) Use of Class II equipment or equivalent insulation
      (iii) Non-conducting location
☐   Absence of protective conductors
      (iv) Earth-free local equipotential bonding
☐   Presence of earth-free equipotential bonding conductors
      (v) Electrical separation (413–06)

☐ 9 Prevention of mutual detrimental influence

☐ 10 Presence of appropriate devices for isolating and switching correctly located

☐ 11 Presence of undervoltage protective devices where appropriate

12 Choice and setting of protective and monitoring devices (for protection against indirect contact and/or overcurrent)

☐   Residual current devices

☐   Overcurrent devices

☐ 13 Labelling of protective devices, switches and terminals

☐ 14 Selection of equipment and protective measures appropriate to external influences

☐ 15 Adequacy of access to switchgear and equipment

☐ 16 Presence of danger notices and other warnings

☐ 17 Presence of diagrams, instructions and necessary information

☐ 18 Erection methods

☐ 19 Requirements of special locations

Tick to indicate item has been inspected
Delete if item not applicable

3 ● Socket A OK as readings are approximately $\frac{1}{2}$ of 0.5 Ω

(3 marks)

● Socket B cross polarity P-cpc, or twisted N conductors not in N terminal   **(3 marks)**

● Socket C loose neutral connection   **(3 marks)**

● Socket D cross polarity P–N, or twisted CPCs not in terminal   **(3 marks)**

● Socket E OK.   **(3 marks)**

4 (a) Conduct the test from the house as this will then include the SWA cable. Disconnect the supply to the PIR controlled lights and the electronic door sensor. Disconnect the capacitor and ballast at each fluorescent luminaire. With the garage main switch and the MCBs ON and any accessories

unplugged, test at 500 V between live conductors connected together and earth and then between each live conductor. Operate the two-way switches during each test. The test readings should not be less than 0.5 MΩ.

(8 marks)

(b) If any reading is below 2 MΩ, then there may be a latent defect and then each circuit should be tested separately and the insulation resistance in each case should be greater than 2 MΩ.

(7 marks)

5 As a loop impedance tester delivers a high current for a short time, it is not unusual for sensitive MCBs with low ratings to trip out on overload. The loop impedance in such cases will have to be determined by a combination of measurement and calculation as follows:

Measure Z_e and measure $(R_1 + R_2)$ for the circuit, then $Z_s = Z_c + (R_1 + R_2)$

(7 marks)

6 (a) ● The point of fault
 ● The cpc
 ● The steel wire armour of the garage supply
 ● The earthing conductor
 ● The metallic earth return path of the supply cable
 ● The earthed neutral of the transformer
 ● The transformer winding
 ● Phase conductors.

(8 marks)

(b) $Z_s = U_{oc}/I$
 $= 240/120$
 $= 2\,\Omega$

so circuit protection will operate fast enough.

(7 marks)

APPENDIX C

The Second Amendment to the IEE Wiring Regulations

The second amendment to the IEE Regulations comprises many editorial changes (which will not significantly affect design, construction or testing procedures), several deletions, many new regulations and a new section on generators. The following text considers the important changes.

Part 2

Definitions

There are several editorial modifications, but the main change is the deletion of any reference to category 1, 2 or 3 circuits. Now, details of segregation of circuits must be obtained from relevant British Standard codes of practice.

There are three new definitions:

1 *Hazardous live part*. A live part which can give under certain conditions of external influence, an electric shock.
2 *Ordinary person*. A person who is neither a skilled person nor an instructed person.
3 *Voltage bands*.
   Band 1: Installations where protection against electric shock is provided under certain conditions by the value of the voltage, and

installations where the voltage is limited for operational reasons (e.g. telecommunications, signalling, bell, control and alarm installations). *Extra low voltage installations Band 2*: Contains the voltages for supplies to household, and most commercial and industrial installations. *Low voltage installations*

Part 3

Compatibility

Reg. 331–01–01: This has been added to and now indicates influences which may influence compatibility with other electrical equipment or other services such as:

- transient overvoltages
- rapidly fluctuating loads
- starting currents
- harmonic currents
- DC feedback
- high-frequency oscillations
- earth leakage currents
- the necessity for additional connections to earth.

Part 4

Protection by SELV

Reg. 411–02–06: There is some new wording in item (ii) regarding SELV conductors, that permits the use of an insulated sheath in addition to basic insulation as a means of electrical separation, rather than physically separating them.

A new paragraph suggests that where SELV and other circuits are contained in equipment such as relays, contactors etc. the degree of separation between live parts of these circuits must be maintained. Manufacturers, designers and installers will have to ensure that equipment meets this requirement.

The type-testing insulation voltage for SELV circuits has been changed from 500 V DC to 500 V AC.

Protection by insulation of live parts

Reg. 412–02–01: This has been added to by suggesting that factory built assemblies be type tested to an appropriate British Standard. It is further suggested that insulation applied on site should be as durable as that applied to factory built assemblies. Paints, lacquers, varnishes are not considered suitable insulation materials.

Earthed equipotential bonding and automatic disconnection of supply

Reg. 413–02–04: This now embodies the content of 413–02–15 and allows, in the case of unacceptable disconnection times by overcurrent devices, for either supplementary bonding to be applied or an RCD installed. This now applies regardless of the earthing system, i.e. TT, TN-S etc.

Tables 4IBI and 41 C Miniature circuit breakers to BS EN 60898 are now referred to as Circuit breakers.

Protection by non-conducting location

Reg. 413–04–07: This has increased the separation distance between exposed and extraneous conductive parts – from 2 m to 2.5 m.

Protection against overload current

Reg. 433–02–04: This relates to ring circuits and the acceptance that not only do 2.5 sq. mm conductors comply but so now do 1.5 sq. mm MI cable conductors (this used to be case in the 15th Edition!).

Reg. 411–02–06 (ii): This has been re-worded to suggest that one means of electrical separation of SELV circuits from other circuits is for the SELV circuit cables to have an insulated sheath additional to the basic insulation.

Part 5

Electromagnetic compatibility
Reg. 515–02:   A new section with two regulations that say what is installed:

(i)   is not affected by surrounding electromagnetic influences, and
(ii) will not cause electromagnetic interference with other equipment.

Presence of water
Reg. 522–03–01: The word humidity has been changed to condensation.

Impact
Reg. 522–06–01:   This has been added to, to include examples of mechanical damage such as impact, compression, tension, abrasion and penetration.

Reg. 522–06–03:   This has been added to, to say that underground cables, unless concentric in nature, must have an earthed armour or sheath that is suitable for a protective conductor and that buried conduits and ducts must be suitably identified.

Reg. 522–06–05 and 07:   This has been added to by allowing the use of insulated concentric cables to provide mechanical protection when installed under floors, above ceilings or in walls or partitions.

Solar radiation
Reg. 522–11–01:   An addition which now includes ultraviolet radiation and ionizing radiation as well a solar radiation.

Building design
Reg. 522–12–03:   This is new and requires that any load-bearing element of a building be such that its purpose will not be impaired by the penetration of a wiring system. (Notching and drilling of joists, for example.)

Neutral conductors

Reg. 524–02–01: This has been changed to consider single-phase circuits only and states that neutral and phase conductors should be the same size.

Reg. 524–02–02: This now considers neutrals in polyphase circuits and suggests that a reduced neutral CSA be justified in terms of likely current, taking into account load balancing, power factors and phase and harmonic currents.

Risk of spread of fire

Reg. 527–01–03: This new regulation indicates that no special precautions need be taken to minimize the spread of fire if cables are installed to BS 4066: Pt. 1. If, however, the risk is high, then special precaution shall be taken.

Reg. 527–01–04: Another new regulation which states that if cables do not comply with BS 4066: Pt. 1, then their use should be restricted to short lengths for connection to appliances to the fixed wiring and should not pass between fire segregated areas or compartments

Reg. 527–01–05: This new regulation deals with resistance to flame propagation by conduit and trunking and suggests that no special precautions need be taken when installed to BS EN 50085:1 or 50086:1

Proximity to electrical services

Reg. 528–01–01: A change of wording here, with ELV and LV circuits being replaced with Band 1 and Band 2 circuits.

Reg. 528–01–02: This regulation now embodies Regs. 528–01–03 to 08 which have been deleted and states that Band 1 and Band 2 circuits must not be contained within the same wiring system unless:

(i)   every cable is insulated to the highest voltage present, or
(ii) for multicore cables, Band 1 cores are insulated individually or collectively to the highest voltage in the Band 2 circuit or

are separated by an earthed metal screen capable of carrying a current equal to that of the largest core of the Band 2 circuit, or

(iii) the cables of each band are run separately in compartmentalized ducting or trunking or on partitioned tray, or

(iv) separate conduit, trunking or ducting systems are used.

When the Band 1 circuits are SELV or PELV, the additional requirements for such systems must also be met.

For telecommunication and similar circuits, the relevant Standard should be referred to.

Residual current devices

Reg. 531–02–09: This is a new regulation which addresses RCDs that are adjustable (quite rare) and suggests that such a device, if adjusted by an ordinary person, be designed and constructed so that any adjustment can only be made by use of a key or tool and that any adjusted setting is clearly visible.

Generators

Regs 551–01–01 to 551–07–05: Twenty-four new regulations dealing with LV or ELV installations that incorporate generators, and outline the requirements for:

(i)   supplies to installations not connected to the public supply

(ii)   supplies to installations as an alternative to the public supply (standby systems)

(iii) supplies to installations in parallel with the public supply

(iv) appropriate combinations of these.

Consideration is given to supplies to:

(i)   permanent installations

(ii)   temporary installations, and

(iii) portable equipment not connected to a permanent fixed installation.

In broad terms this new section considers:

(i) ELV systems supplied from more than one source where the relevant regulations in Part 4 for SELV, PELV and FELV apply

(ii) protection against indirect contact by automatic disconnection of supply, where Reg. 413–02 applies

(iii) additional requirements to (ii) above for standby systems, static inverters, and portable generating sets supplying installations that are not permanently fixed

(iv) protection against overcurrent

(v) additional requirements for standby systems to prevent parallel operation with the public supply

(vi) additional requirements where a generator may be operated in parallel with the public supply.

Part 6

Hot air saunas
Reg. 603–7–01:   A change in the temperature rating of BS 6141 flexible cords from 15°C to 180°C.

Caravans
Reg. 608–03–02:   This has been amended to require RCDs to break all live conductors.

Part 7

Testing
Reg. 713–01–01:   This regulation has been rearranged and added to in so far as there is now a requirement to compare test results with relevant criteria such as design values or tabulated values in BS7671. There is also a requirement to conduct the first eight tests on a new installation before it is energized.

Initial verification

Regs 742–01–01 and 02: The only changes here are the re-
naming of 'Completion Certificate' to 'Electrical Installation
Certificate', and that such certificates may be authenticated other
than by signature.

Alterations and additions

Reg. 743–01–02: This introduces the use of a Minor Works
Certificate for installation work that does not make provision for a
new circuit. Such a certificate must be signed or otherwise
authenticated by a competent person, stating that the work has
been designed, constructed and inspected and tested in accordance
with the Regulations.

Appendix 6

This give examples of all the new certificates.

APPENDIX D

The Third Amendment to the IEE Wiring Regulations

Section 601: Locations containing a bath or shower

This amended section deals with common locations containing baths, showers and cabinets containing a shower and/or bath. It does not apply to specialist locations. The main feature of this section is the division of the location into zones (0, 1, 2 and 3) in the same way as section 602 for swimming pools.

Zone 0

This is the interior of the bath tub or shower basin or, in the case of a shower area without a tray, it is the space having a depth of 500 mm above the floor out to a radius of 600 mm from a fixed shower head or 1200 mm radius for a demountable head.

Figure D1

Points to note

- Only SELV (12 V) may be used as a measure against electric shock, the safety source being outside zones 0, 1 and 2.
- Other than current using equipment specifically designed for use in this zone, **no** switchgear or accessories are permitted.
- Equipment designed for use in this zone must be to at least IP X7.
- Only wiring associated with equipment in this zone may be installed.

Zone 1

This extends above zone 0 around the perimeter of the bath or shower basin to 2.25 m above the floor level, and includes any space below the bath or basin that is accessible without the use of a key or tool. For showers without basins, zone 1 extends out to a radius of 600 mm from a fixed shower head or 1200 mm radius for a demountable head.

Points to note

- Only SELV may be used as a measure against electric shock, the safety source being outside zones 0, 1 and 2.

Figure D2

- Other than switches and controls of equipment specifically designed for use in this zone, and cord operated switches, only SELV switches are permitted.
- Provided they are suitable, fixed items of current using equipment that may be installed in this zone are water heaters and shower pumps.
- Other equipment that can reasonably be only installed in this zone must be suitable, and be protected by an r.c.d. of rated tripping current 30 mA or less.
- Equipment designed for use in this zone must be to at least IP X4, or IP X5 where water jets are likely to be used for cleaning purposes.
- Only wiring associated with equipment in this zone and zone 0 may be installed.

Zone 2

This extends 600 mm beyond zone 1 and to a height of 2.25 above floor level and to the space above zone 1 between 2.25 m and 3 m above floor level.

Points to note

- Only SELV (12 V) may be used as a measure against electric shock, the safety source being outside zones 0, 1 and 2.

Figure D3

- Other than current using equipment specifically designed for use in this zone, only switches and socket outlets of SELV circuits and shaver units to BS EN 60742 may be installed.
- Provided they are suitable for use in this zone, water heaters, shower pumps, luminaires, fans etc. may be installed.
- Equipment designed for use in this zone must be to at least IP X4, or IP X5 where water jets are likely to be used for cleaning purposes.
- Only wiring associated with equipment in this zone and zones 0 and 1 may be installed.

Zone 3

This extends 2.4 m beyond zone 2 and to a height of 2.25 m above floor level and to the space above zone 2 between 2.25 m and 3 m above floor level.

Points to note

- Only SELV socket outlets or shaver units to BS EN 60742 may be installed.
- Equipment other than that specifically designed for use in this zone, must be protected by an r.c.d. of rated tripping current 30 mA or less.
- Equipment designed for use in this zone must be to at least IP X5 if water jets are likely to be used for cleaning purposes.

Figure D4

Appendix D

Supplementary equipotential bonding

Supplementary bonding must be carried out and will connect together the c.p.c. terminals of Class I and Class II equipment in zones 1, 2 and 3 and any extraneous conductive parts within these zones.

Such extraneous conductive parts will include:

● metallic gas, water, waste and central heating pipes
● metallic structural parts that are accessible to touch
● metal baths and shower basins.

This bonding may be carried out in the bath or shower room or in close proximity to it.

Electric floor units may be installed below any zone provided that they are covered with an earthed metal grid or metallic sheath and connected to the local supplementary bonding.

Where a cabinet or cubicle containing a bath or shower is installed in a room other than a bathroom, any socket outlet must either be a SELV or shaver socket or be protected by an r.c.d. of rated tripping current 30 mA or less. Also, local supplementary bonding is not required.

IEE Wiring Regulations

The tables referred to in this book appear in the *IEE Wiring Regulations* and the *On-Site Guide*. These publications and other guidance notes can be obtained direct from the Institution of Incorporated Engineers (IEE), Savoy Place, London WC2R 0BL. Tel. 020 7240 1871.

World Wide Web URL:

   http://www.iee.org.uk/publish/books/wireregs.html

INDEX

Index